SOLDE

HOW TO SOLDER FOR BEGINNERS

LUKE H.

CONTENTS

CHAPTER ONE

INTRODUCTION

Soldering is a type of metal joining process that involves melting a filler material (solder) between two surfaces to create an electrical or structural bond. It is one of the most common methods of creating electrical connections and is widely used in the electronics and electrical industries. Soldering is used in a variety of applications, from repairing and assembling electronics to joining plumbing pipes. With the right tools and techniques, soldering can be a safe and effective way to join metals together. Soldering involves heating the parts to be joined to the point where the solder melts and flows into the joint. These make a strong bond between the two pieces. The process requires using the correct tools and materials, following safety guidelines, and understanding the steps involved in the process. Once you have the necessary materials and tools, the process of soldering can be divided into five steps: preparation, heating, fluxing, soldering, and cleaning. Preparation involves cleaning the surfaces to be joined and pre-tinning the wires. Heating involves using a soldering iron or torch to heat the joint. Fluxing is the process of applying a fluxing agent to the joint to promote good adhesion. Soldering involves melting the solder and flowing it into the joint. Finally, cleaning involves removing any excess solder and flux residue. By following the steps outlined above, you can successfully solder two pieces of metal together. Soldering is a skill that takes practice to master, but with the right tools, safety precautions, and knowledge of the process, you can become an expert solder in no time.

CHAPTER TWO

TYPES OF SOLDERING IRONS

1. Corded Soldering Irons - These are the most common type of soldering iron. They are usually powered by electricity, and require a cord to be plugged into a power source in order to work.

2. Cordless Soldering Irons - These are similar to corded soldering irons, but they are powered by rechargeable batteries instead. They are usually more portable and convenient, but have shorter battery

life and may be more expensive than corded soldering irons.

3. Solder Pots - These are large tanks that are filled with a liquid or solid form of solder. The heat is applied to the solder, which then melts and is used to join pieces together.

4. Solder Guns - These are tools that use a heated metal tip to melt solder quickly and accurately. They are usually used for larger projects and are often used by professional electricians.

5. Solder Stations - These are more advanced soldering tools that use a combination of temperature control and other features to make soldering

easier and more precise. They are often used by electronics professionals and hobbyists.

6. Hot Air Soldering Stations - These are specialized tools that use a stream of hot air to melt the solder. They are often used for more complicated soldering projects and are often found in electronics repair shops.

7. Solder Fumes Extractors - These are specialized tools that are used to remove hazardous fumes that can be produced when soldering. They are often found in electronics repair shops.

8. Solder Paste - This is a type of solder that comes in a paste form. It is often used for surface mount projects, and is considered to be a safer alternative to traditional soldering methods.

9. Solder Wire - This is a type of solder that comes in a spool of wire. It is often used for through-hole projects, and is considered to be a safer alternative to traditional soldering methods.

10. Solder Wick - This is a type of solder that comes in a coil of wicking material. It is used to remove excess solder from a joint.

11. Desoldering Tools - These are specialized tools that are used to remove solder from a joint. They are often used in electronics repair shops.

12. Solder Flux - This is a type of chemical that is used to help the solder flow more easily. It is often used in combination with solder wire or solder paste.

13. Solder Preforms - These are small pieces of solder that come in a variety of shapes and sizes. They are often used in combination with solder wire or solder paste.

14. Solder Tips - These are attachments that are used to hold the

solder and direct the heat. They come in a variety of shapes and sizes, and are often used in combination with other soldering tools.

15. Solder Baths - These are specialized baths that are filled with liquid solder. They are often used for larger projects and are often found in electronics repair shops.

16. Solder Fountains - These are specialized tools that are used to dispense a steady stream of solder. They are often found in electronics repair shops.

CHAPTER THREE

STANDARD SOLDERING IRONS

Standard soldering irons are basic, low-cost tools used for a variety of electrical and electronics projects. These irons usually have a metal shaft with a handle and a flat, conical tip at the end. They are powered by electric current and are designed to heat up to a certain temperature. Standard soldering irons are often used to solder circuit boards, components and wires together as part of a larger electrical project. They are also used to cut, shape and form metal for a variety of applications. When used correctly, standard soldering irons can provide a safe and reliable way to make electrical connections. However, improper use of the tool can cause damage and even fire hazards. Therefore, it is important to use the correct soldering iron for the task at hand and to follow all safety precautions when using the tool. Overall, standard soldering irons are an economical and useful tool for a variety of electrical and electronic applications. They are relatively easy to use and provide a reliable way to make electrical connections safely.

In the end, the type of soldering iron you choose will depend on the specific task you are trying to accomplish. Make sure to do your research and select the right tool for the job.

Temperature-Controlled Soldering Irons

Temperature-controlled soldering irons are essential tools for all electronics enthusiasts. They provide precision control over the temperature of the soldering tip, which is necessary for delicate, intricate soldering jobs. Temperature-controlled soldering irons are equipped with a digital readout that allows the user to monitor and adjust the temperature of the soldering tip. They also offer a wide range of tips and other accessories, allowing the user to tackle a variety of soldering tasks. Temperature-controlled soldering irons are typically more expensive than standard soldering irons, but they offer far greater control and accuracy. They also often last longer than other types of soldering irons. If you are serious about your electronics projects, investing in a quality temperature-controlled soldering iron is a must.

Save yourself time, money, and frustration with a temperature-controlled soldering iron. It will enable you to complete complex soldering tasks quickly, accurately, and safely.

Gas-Powered Soldering Irons

Gas-powered soldering irons are a type of soldering tool that use a gas-powered flame to heat up the tip of the soldering iron. These tools are typically used for plumbing, welding, and other tasks that require intense heat and precision. They are usually made of either brass or stainless steel and use a propane or butane fuel source. The flame produced by the fuel can reach temperatures of up to 2500°F, allowing for a more precise and controlled heat than an electric soldering iron. Gas-powered soldering irons are popular among professionals, as they are more powerful and provide more control than electric ones. They are especially useful for soldering large objects or projects with intricate details. However, they are more expensive than electric soldering irons, and require regular maintenance and fuel refills. Gas-powered soldering irons can also be dangerous if not used properly, so it is important to read the instructions thoroughly before use. Overall, gas-powered soldering irons are a great tool for those who need higher temperatures and more precise control than what electric soldering irons can provide. They are expensive, however, and require regular maintenance and fuel refills. As such, it is important to weigh the pros and cons before purchasing a gas-powered soldering iron.

CHAPTER FOUR

CHOOSING THE RIGHT SOLDERING IRON

When selecting a soldering iron, it is important to consider the type of soldering job you will be doing. Different soldering irons have different features that make them suitable for different applications. Some types of soldering irons are better suited for delicate work, while others are better for more heavy-duty work. Additionally, the wattage of the iron should be taken into consideration, as a higher wattage will allow for more heat to be applied to the connection points. Finally, it is important to make sure that the iron has the right tip size and type for your specific purpose, as having the wrong tip size can lead to poor connections. Overall, when choosing a soldering iron it is important to consider the type of work you will be doing, the wattage of the iron, and the tip size and type. By taking these factors into consideration, you can ensure you select the right tool for the job. In conclusion, when selecting a soldering iron, it is important to consider the type of work you will be doing, the wattage of the iron, and the tip size and type. This will help ensure that you have chosen the right tool for the job, and will help you to make quality connections.

Consider the Type of Work

If you are looking for a soldering iron for electrical work, you should look for a soldering iron that is rated for high temperatures, has a long cord for greater reach, and is insulated for safety. It should also come with a variety of tips for different types of soldering work, such as electronics, plumbing, and jewelry. Additionally, a soldering iron should come with a stand for safe storage and a sponge for cleaning the tip. If you are looking for a soldering iron for jewelry work, you should look for a soldering iron specifically designed for the job. These soldering irons usually have a lower temperature and a narrow tip, which is ideal for delicate soldering work. A soldering iron designed for jewelry work should also come with a stand, sponge, and a variety of tips for different types of jewelry work. Finally, if you are looking for a soldering iron for hobby work, you should look for a soldering iron that is relatively inexpensive and easy to use. These soldering irons usually have a lower temperature and a wide tip, which is ideal for hobby projects. A soldering iron for hobby work should also come with a stand and a sponge for cleaning the tip.

Regardless of the type of work you are doing, it is important to make sure that you purchase a soldering iron with the necessary safety features and that is suitable for the task.

Consider the Power Source

The power source for a soldering iron is typically an AC or DC power supply. AC power sources are most common, as AC power is easier to regulate and often more available than DC power. However, some soldering irons can be powered by batteries or solar cells. In addition to the power source, the wattage of the soldering iron is also important.

Generally, higher wattage models can provide more heat, allowing for faster and more efficient soldering. However, too much heat can damage components, so it's important to choose the right wattage for your project. Finally, some soldering irons offer adjustable temperature control, allowing you to customize the heat output for a particular application. Overall, the power source for a soldering iron is an important consideration when selecting the right tool for the job. Make sure to do your research to find the power source and wattage that best suits your application.

Consider the Tip Type

1. Conical Tip: The conical tip is the most common and popular tip shape for soldering irons. It is usually used for soldering components on Printed Circuit Boards (PCBs) and other small components.

2. Chisel Tip: The chisel tip is a wider and flatter tip shape that is used for soldering larger components, such as connectors and switches. It is also suitable for soldering larger areas.

3. Knife Tip: The knife tip is a narrower, more pointed tip shape that is used for precision soldering. It is useful for soldering small surface-

mount components, such as resistors and capacitors.

4. Screwdriver Tip: The screwdriver tip is a specialized tip shape that is used for soldering terminals and wires. It is also useful for removing components from PCBs.

5. Spade Tip: The spade tip is a flat, broad tip shape that is used for soldering large areas quickly. It is especially useful for soldering large components, such as heat sinks and connectors.

6. Bevel Tip: The bevel tip is a specialized tip shape that is used for soldering through-hole components.

It is especially useful for soldering components on double-sided PCBs.

7. Pyramid Tip: The pyramid tip is a triangular tip shape that is used for soldering components in tight places. It is especially useful for soldering components on crowded PCBs.

8. Loop Tip: The loop tip is a circular tip shape that is used for soldering wires and terminals. It is especially useful for soldering insulated wires.

9. Bent Tip: The bent tip is a specialized tip shape that is used for soldering in hard-to-reach places. It is especially useful for soldering components in tight spaces.

10. Specialty Tip: Specialty tips are custom-made tips that are designed for specific soldering applications. These tips are usually designed for a specific component or application and are not available in the standard tip shapes.

11. Solder Sucker Tip: The solder sucker tip is a specialized tip shape that is used for desoldering components. It is especially useful for removing components from PCBs.

Consider Additional Features for soldering iron

1. Automatic Temperature Control: This feature helps maintain a consistent temperature for better soldering.

2. Adjustable Tip Sizes: This feature allows for different tip sizes to be used for different soldering jobs, making it more versatile.

3. LED Lighting: This feature helps provide better visibility for soldering in dark areas.

4. Heat Protection: This feature helps protect your work station from the heat generated by the soldering iron.

5. Fast Heat-up Time: This feature helps reduce the time taken for the iron to reach the desired temperature.

6. Ergonomic Design: This feature helps make the soldering iron comfortable to use and reduce fatigue.

7. Anti-Static Protection: This feature helps prevent static electricity from damaging sensitive components.

8. Replaceable Tips: This feature allows for the tips of the soldering iron to be replaced when they become worn or damaged.

9. Stand or Holder: This feature helps keep the soldering iron in a safe place while not in use.

10. Safety Shut-off: This feature helps prevent the soldering iron from being used for too long and overheating.

11. Digital Display: This feature helps provide more accurate temperature and power settings for soldering.

12. Auto-shut Off: This feature helps prevent accidental overheating of the soldering iron.

13. Portable Design: This feature helps make the soldering iron more convenient to use in different places.

14. Interchangeable Power Sources: This feature allows the soldering iron to be powered by different sources, such as AC, DC, and battery.

15. Variable Heat Settings: This feature allows for more precise control of the heat of the soldering iron.

16. Auto Sleep Mode: This feature helps conserve energy when the soldering iron is not in use.

17. Auto Cool Down: This feature helps cool down the soldering iron quickly after use.

18. Heat Insulation Handle: This feature helps protect the user's hands

from the heat generated by the soldering iron.

19. Magnetic Tool Holder: This feature helps keep the soldering iron in a secure place while not in use.

20. ESD (Electrostatic Discharge) Protection: This feature helps protect sensitive components from static electricity.

21. Solder Saver: This feature helps conserve solder by allowing it to be reused.

22. Solder Retrieval: This feature helps retrieve solder that has been spilled.

23. Solder Dispenser: This feature helps make soldering faster and easier by dispensing the solder directly onto the soldering iron tip.

24. Solder Tip Cleaner: This feature helps keep the soldering iron tip clean for better performance.

25. Solder Wick: This feature helps remove excess solder from components.

26. Vacuum Extractor: This feature helps remove flux, smoke, and other particles from the soldering area.

CHAPTER FIVE

SAFETY TIPS FOR SOLDERING

1. Always wear safety glasses when soldering to protect your eyes from the bright light and sparks that come from the soldering iron.

2. Use a damp sponge to clean the soldering iron tip as you work, as this will help to prevent oxidation of the tip and will keep the iron operating at a safe temperature.

3. Before plugging your soldering iron into an electrical outlet, ensure that the voltage and amperage

ratings of the iron match the voltage of the outlet.

4. Make sure to unplug the soldering iron when you are finished working, and never leave it unattended while it is still plugged in.

5. Be aware of the temperature of the soldering iron and never touch it with your bare hands.

6. Always keep your work area clean and free from clutter, and make sure to keep flammable materials away from the soldering iron.

7. Keep soldering wires away from any other electrical wires to prevent short-circuiting.

8. Take breaks from soldering every 20-30 minutes to give your eyes a rest from the bright light.

9. If you are working on a project that requires a lot of soldering, it is strongly recommended that you use a fume extractor to reduce your exposure to the toxic fumes produced by the soldering process.

10. If you are a beginner, it is always a good idea to have an experienced solderer nearby to help you out if needed.

CHAPTER SIX

CONCLUSION

A soldering iron is a great tool for making repairs and creating electrical connections. It is an essential tool for any electrical work, from simple soldering jobs to complex circuit board assembly. With the right tools and techniques, soldering can be a safe and rewarding experience. With care and practice, any user can become proficient in soldering and reap the rewards of a well-made electrical connection.

Overall, a soldering iron is an important tool for anyone who works on electronics or electrical connections. Its versatility and safety make it one of the most important tools in the electrical repair and assembly toolbox. I hope this book has helped you learn more about soldering irons and their uses. Best of luck in your soldering endeavors!

Printed in Great Britain
by Amazon